"Uncovering the Surprising Facts that Will Change Your Worldview"

"Unlocking the Power of Knowledge: Discover the Mind-Blowing Truths That Challenge Everything You Thought You Knew"

With:
Abubakar Sunusi Usman

Copyright©Abubakar Sunusi Usman2023

All right reserved

Table of contents

INTRODUCTION
THE HUMAN MIND
HISTORY
SCIENCE
POLITICS
SOCIETY
ENVIRONMENT
ECONOMICS
TECHNOLOGY
CONCLUSION

INTRODUCTION

- Worldview: A worldview refers to an individual's fundamental beliefs and assumptions about the world, including their values, attitudes, and cultural background. It is a lens through which they interpret and understand the world around them.

- Assumptions: Assumptions are beliefs that are taken for granted without questioning their validity or accuracy. They are often based on past experiences, cultural norms, and personal biases. Assumptions can limit our thinking and prevent us from considering alternative perspectives.

- Perspectives: Perspectives refer to a person's point of view, or how they see and interpret the world around them. Perspectives can be influenced by factors such as culture, upbringing, education, and personal experiences. Having a broad perspective means being open to different ideas and considering multiple viewpoints.

- Challenging assumptions: Challenging assumptions means questioning the validity and accuracy of our beliefs and assumptions. It involves being open to new information and different perspectives and being willing to revise our thinking based on new evidence.

- Broadening perspectives: Broadening perspectives means expanding our understanding of the world by considering different viewpoints and experiences. It involves being curious and open-minded and seeking out new information and perspectives.

•Surprising facts: Surprising facts are pieces of information that challenge our assumptions and broaden our perspectives. They can be unexpected, counterintuitive, or even shocking, but they provide us with new insights and opportunities to learn and grow.

In summary, challenging assumptions and broadening perspectives are important for developing a more nuanced and accurate worldview. Surprising facts can help us question our assumptions and expand our perspectives, leading to greater understanding and personal growth.

THE HUMAN MIND

The human mind is a complex system that is capable of processing vast amounts of information and performing a wide range of cognitive functions. At the same time, it is subject to a number of limits and biases that can affect our perception and decision-making processes.

•Cognitive biases are inherent tendencies in human thinking that can lead to errors and distortions in our judgment. For example, confirmation bias is the tendency to seek out and interpret information in a way that confirms our pre-existing beliefs or biases, while availability bias is the tendency to rely on easily available or memorable information rather than considering all available evidence.

•Psychology is the scientific study of the mind and behavior. It encompasses a wide range of topics, including perception, cognition, emotion, motivation, personality, social behavior, and mental health. Through psychological research, we can gain a better understanding of how the mind works and how we can optimize its potential.

•Perception refers to the process by which we interpret sensory information from the environment. Our perception is shaped by a range of factors, including our past experiences, expectations, and cultural background. For example, optical illusions demonstrate how our perception can be easily fooled by visual cues.

•Consciousness is the state of being aware of one's surroundings and having subjective experiences, thoughts, and emotions. The nature of consciousness is a topic of much debate and research, with various theories attempting to explain its underlying mechanisms.

Overall, the human mind is a complex and fascinating system that has both limits and possibilities. By understanding its workings and potential biases, we can improve our decision-making, problem-solving, and overall cognitive performance.

HISTORY

History is the study of past events, people, and societies. However, the interpretation and presentation of history can vary widely depending on the perspectives of those who tell it. This has led to the emergence of revisionism, historical myths, alternative narratives, and forgotten events.

•Revisionism refers to the re-examination of historical events and the questioning of traditional interpretations. This approach challenges established narratives and seeks to reinterpret events through new evidence, perspectives, and historical contexts. For example, a revisionist view of World War II might challenge the traditional portrayal of the war as a simple struggle between good and evil, instead examining the complex motivations and interests of different nations and groups involved.

•Historical myths are beliefs or stories about the past that are not based on historical fact, but rather on popular misconceptions, propaganda, or wishful thinking. These myths can be perpetuated by individuals, groups, or societies, and may serve political or cultural agendas. For example, the myth of the American Wild West portrays cowboys as heroic figures who fought against evil outlaws, but this myth ignores the reality of racism, violence, and exploitation that existed during that time

.•Alternative narratives are accounts of history that diverge from the dominant or official version. These narratives may arise from marginalized or minority perspectives that have been overlooked or suppressed by mainstream history. Alternative narratives can shed light on previously ignored or silenced voices, and offer a more nuanced understanding of historical events. For example, the history of the civil rights movement in the United States is often told from the perspective of black activists, who challenged the dominant narrative of white heroism.

•Forgotten events refer to historical events that have been overlooked or forgotten in the dominant narrative of history. These events may be small-scale or large-scale, and can represent significant moments in the lives of individuals or groups. For example, the story of the Tuskegee Airmen, a group of African American pilots who fought in World War II, was largely forgotten until the 1990s, when their contributions were finally recognized and celebrated.

In conclusion, revisiting the past can involve challenging established narratives through revisionism, uncovering and examining historical myths, exploring alternative narratives that offer different perspectives, and recovering forgotten events that have been overlooked or silenced by mainstream history. This approach to history can help us to develop a more complex and nuanced understanding of the past, and to appreciate the diversity of human experience.

SCIENCE

Science is a complex and ever-evolving field of study that seeks to understand the natural world through observation, experimentation, and analysis. While much of the knowledge we have about the world around us is based on commonly accepted theories and empirical evidence, there are many scientific discoveries, unconventional theories, paradigm shifts, and mysteries of the universe that lie beyond common knowledge.

•Scientific discoveries refer to breakthroughs in understanding that challenge or expand upon existing knowledge. These can include new technologies, medicines, and other innovations, as well as insights into the workings of the natural world, such as the discovery of new species, geological formations, or astronomical phenomena.

•Unconventional theories, on the other hand, are ideas that challenge widely accepted scientific principles and theories. These may be initially dismissed or ridiculed by the scientific community, but if they are supported by empirical evidence and rigorous testing, they can eventually gain acceptance and become part of mainstream science.

•Paradigm shifts are transformative moments in scientific understanding that fundamentally alter the way we think about the world. These can occur when new evidence challenges old assumptions or when new technologies enable us to observe the world in new ways. Examples of paradigm shifts in science include the discovery of the heliocentric model of the solar system, the theory of evolution by natural selection, and the development of quantum mechanics.

•Finally, the mysteries of the universe are those phenomena that we do not yet fully understand, despite our best efforts to study and explain them. These can include dark matter, dark energy, the origin of life, and the nature of consciousness, among others. Scientists continue to explore these mysteries through a combination of observation, experimentation, and theoretical modeling, in the hopes of one day uncovering their underlying principles and mechanisms.

Overall, science is a constantly evolving field that seeks to expand our understanding of the world around us, and while much of what we know is based on commonly accepted knowledge, there is always more to discover and explore.

POLITICS

Politics can be described as the process of making decisions that affect a society or a community. It involves the exercise of power and influence to shape policies, laws, and regulations that govern how people live and interact with one another. Politics is often associated with power, corruption, and manipulation because those who hold political power can use it to their advantage, sometimes at the expense of others.

- In a democratic system, politics is characterized by free and fair elections, with power being vested in elected officials who are accountable to the people. However, even in democracies, there can be instances of corruption and manipulation, as some politicians seek to use their position for personal gain or to maintain their hold on power.

- In authoritarian systems, politics is often characterized by a lack of political freedoms and the absence of meaningful participation by citizens in decision-making. Political power is often concentrated in the hands of a small group of elites who use it to maintain their grip on power and suppress dissent.

•Propaganda and social control are often used in politics to influence public opinion and shape the narrative around certain policies or issues. Propaganda can be used to manipulate people's emotions and beliefs, often by disseminating false or misleading information. Social control can involve a range of tactics, including censorship, surveillance, and the use of force, to maintain order and suppress dissent.

Overall, politics is a complex and multifaceted field that touches on many aspects of human life. While it can be associated with negative aspects such as power, corruption, and manipulation, it also has the potential to bring about positive change and improve people's lives when practiced with integrity and accountability.

SOCIETY

Society is a complex and diverse system made up of people from different cultural backgrounds, with varying beliefs, values, and behaviors. It is a web of relationships and interactions that shape the way we live and interact with each other.

•One of the defining features of society is its diversity, which encompasses differences in race, ethnicity, gender, sexual orientation, religion, socioeconomic status, and many other factors. This diversity enriches society by exposing people to different perspectives and experiences, but it can also create tensions and conflicts.

•Inequality is a pervasive issue in society, with some individuals and groups enjoying greater access to resources, opportunities, and privileges than others. This inequality can be due to systemic factors such as discrimination and bias, as well as individual choices and circumstances.

•Social issues such as poverty, crime, education, healthcare, and climate change also play a significant role in shaping society. These issues can impact individuals and communities in different ways, with some groups bearing a disproportionate burden.

•Human behavior is another key aspect of society, and it is influenced by a complex interplay of biological, psychological, and social factors. Understanding human behavior is essential for addressing social issues and promoting positive social change.

Overall, society is a complex and dynamic system that requires ongoing efforts to address diversity, inequality, and social issues while fostering positive human behavior and interactions.

ENVIRONMENT

The Earth is a complex and dynamic system that supports a wide array of life forms, including humans. However, it is facing a number of significant challenges that threaten the health and well-being of both the planet and its inhabitants. Some of the most pressing issues related to the environment include climate change, ecology, sustainability, environmentalism, and biodiversity.

•Climate change is the long-term alteration of global weather patterns caused by human activities, such as burning fossil fuels and deforestation, which release large amounts of greenhouse gasses into the atmosphere. These gasses trap heat and contribute to rising global temperatures, leading to more extreme weather events, sea level rise, and other negative impacts.

•Ecology is the study of how organisms interact with each other and their environment. This includes understanding how ecosystems function, the role of biodiversity in maintaining ecosystem health, and the impact of human activities on these systems. Human activities such as habitat destruction, pollution, and climate change are causing significant disruptions to ecosystems and threatening the survival of many species.

•Sustainability refers to the ability of natural systems to continue functioning over time, without being depleted or degraded. This includes maintaining the health of ecosystems and ensuring that the resources we rely on, such as water and energy, are used in a way that can be sustained for future generations. Achieving sustainability requires a shift towards more responsible and efficient resource use, as well as reducing the impact of human activities on the environment.

- Environmentalism is a social and political movement that advocates for the protection of the natural environment and the sustainable use of natural resources. It involves raising awareness about environmental issues, advocating for policy change, and promoting sustainable lifestyles.

- Biodiversity refers to the variety of life forms that exist on Earth, including plants, animals, fungi, and microorganisms. Biodiversity is essential for maintaining healthy ecosystems, providing ecosystem services such as clean air and water, and supporting human well-being. However, human activities are causing significant declines in biodiversity, with many species facing extinction due to habitat destruction, pollution, and climate change.

Overall, addressing these environmental challenges requires a concerted effort from individuals, communities, businesses, and governments around the world. This includes adopting sustainable practices, reducing greenhouse gas emissions, protecting and restoring ecosystems, and promoting biodiversity conservation.

ECONOMICS

Economics is a social science that studies how societies allocate scarce resources among alternative uses, with the aim of maximizing welfare. Wealth and poverty are central to the study of economics, as they reflect the distribution of resources within a society. The study of economics examines the hidden realities that influence economic behavior, including social norms, culture, and institutions.

•One of the main concerns of economics is global inequality, which refers to the unequal distribution of resources and opportunities across countries and regions. This is a complex issue that is influenced by a variety of factors, including historical legacies, political institutions, and market forces.

•Resource distribution is another important topic in economics, as it determines how goods and services are produced and distributed within a society. This includes the allocation of natural resources, such as land and water, as well as human resources, such as labor and skills.

•Market forces are a key concept in economics, as they are the mechanism through which resources are allocated in a market economy. Prices are determined by the interaction of supply and demand, and they play a critical role in guiding economic decisions and resource allocation.

•Financial systems are also an important area of study in economics, as they are the means through which savings are channeled into investment and economic growth. The functioning of financial systems is influenced by a variety of factors, including government regulations, market forces, and technological innovations.

Overall, economics is a broad and complex field that encompasses many different topics and perspectives. By studying the distribution of wealth and poverty, resource allocation, market forces, and financial systems, economists seek to understand the workings of the global economy and identify ways to promote greater prosperity and welfare for all.

TECHNOLOGY

Technology refers to the application of scientific knowledge for practical purposes, especially in industry. It encompasses a broad range of tools, techniques, and systems that facilitate the production, exchange, and consumption of goods and services.

•Advancements in technology refer to the development and improvement of new tools, techniques, and systems that enhance productivity, efficiency, and convenience. These advancements can be seen in various fields, including communication, healthcare, transportation, and entertainment.

- However, along with advancements come risks. Technology can be misused, leading to unintended consequences that may negatively impact individuals, society, and the environment. For example, the rise of social media has led to concerns about online privacy, cyberbullying, and the spread of fake news.

- Innovation refers to the creation of new ideas, products, or services that provide value to users. It is an essential part of technological advancements, driving progress and competition.

- Automation refers to the use of machines or software to perform tasks that were previously done by humans. It has revolutionized various industries, such as manufacturing, logistics, and finance. However, automation also raises concerns about job displacement and the need for re-skilling.

•Artificial Intelligence (AI) is a branch of computer science that focuses on the creation of intelligent machines that can perform tasks that typically require human intelligence, such as problem-solving, decision-making, and natural language processing. AI has the potential to transform various industries, but it also raises concerns about ethical considerations, such as bias, transparency, and accountability.

•Ethics refers to the principles of conduct that govern the behavior of individuals and organizations. In the context of technology, it involves making decisions that prioritize the well-being of individuals and society while also considering the impact on the environment. Ethical considerations are essential in ensuring that technological advancements benefit everyone and do not lead to harm or exploitation.

CONCLUSION

"Embracing the Unknown" is a call to action to be open-minded, curious, and willing to explore the unknown. This means being willing to question our assumptions and preconceived notions, and to approach new ideas and perspectives with intellectual humility. It also means being willing to engage in critical thinking, evaluating evidence and arguments carefully before coming to conclusions.

•Embracing the unknown can lead to personal growth and development, as we expose ourselves to new experiences and ideas. It can also help us build stronger relationships, as we learn to appreciate and understand others who may have different perspectives or experiences from our own.

To fully embrace the unknown, we need to cultivate a mindset of openness, curiosity, and intellectual humility. We should be willing to explore new ideas and experiences, even if they challenge our existing beliefs. We should also be willing to acknowledge when we are wrong or when our beliefs are based on flawed assumptions.

Ultimately, embracing the unknown requires us to let go of our fear of the unfamiliar and to embrace the possibilities that come with exploring new horizons. By doing so, we can become more open, empathetic, and thoughtful individuals, capable of contributing to a more vibrant and inclusive world.

•Open-mindedness involves a willingness to consider new ideas and perspectives, even if they conflict with one's existing beliefs. Curiosity is the desire to explore and learn about new things, asking questions and seeking answers. Critical thinking involves evaluating information and evidence in a logical and rational manner, without being swayed by emotional or ideological biases. Intellectual humility involves recognizing the limits of one's knowledge and being willing to accept when one is wrong or uncertain.

•Curiosity is the drive to seek out new knowledge, experiences, and ideas. It involves a willingness to explore the unknown, to ask questions, and to challenge assumptions. Curiosity is a fundamental aspect of learning and discovery, and it can lead to personal growth and a deeper understanding of the world around us.

•Critical thinking is the ability to analyze and evaluate information, arguments, and evidence in a thoughtful and reasoned way. It involves questioning assumptions, assessing the reliability of sources, and considering different perspectives. Critical thinking is important for making informed decisions, solving complex problems, and engaging in meaningful discussions.

•Intellectual humility is the recognition that one's own knowledge and beliefs are limited and fallible. It involves a willingness to listen to others, to consider alternative perspectives, and to revise one's own beliefs in the face of new evidence. Intellectual humility is important for fostering open-mindedness, respectful dialogue, and a willingness to learn from others.

•Personal growth is the process of developing one's own potential, talents, and abilities. It involves setting goals, seeking out new experiences, and continuously learning and improving oneself. Personal growth can be facilitated by curiosity, critical thinking, and intellectual humility, as these qualities encourage individuals to challenge themselves, to seek out new knowledge and experiences, and to reflect on their own beliefs and assumptions. Personal growth can lead to greater self-awareness, improved relationships, and a deeper sense of purpose and fulfillment in life.

In conclusion, "Uncovering the Surprising Facts that Will Change Your Worldview" is a thought-provoking and enlightening book that challenges readers to question their assumptions and beliefs. By presenting a wealth of surprising and sometimes unsettling facts, the book forces readers to confront uncomfortable truths about the world around them. However, while the information may be unsettling at times, the book ultimately offers a message of hope by showing how individuals can take action to make a positive difference in the world. For those willing to challenge their own worldview and explore new perspectives, "Uncovering the Surprising Facts that Will Change Your Worldview" is a must-read that has the potential to transform not just individual perspectives, but also the world at large. Overall, the book offers a valuable contribution to the ongoing dialogue about the issues that face our world today, and is sure to leave a lasting impact on readers who engage with its powerful message.

ABUBAKAR SUNUSI USMAN

www.ingramcontent.com/pod-product-compliance
Lightning Source LLC
Chambersburg PA
CBHW030518220526
45464CB00006B/2847